THIRD EYE AWAKENING

The Ultimate Beginner's Guide to Opening Your Third Eye Chakra

Chloe Brisbane

© **Copyright 2018 by Chloe Brisbane - All rights reserved.**

No part of this book may be reproduced or transmitted in any form or by any means, electronic or mechanical, including photocopying, recording or by any information storage and retrieval system without written permission of the publisher, except for the inclusion of brief quotations in a review.

TABLE OF CONTENTS

INTRODUCTION 1

Chapter 1 *See The Light* 4

Chapter 2 *Restoring Your Soul* 17

Chapter 3 *Lifting The Veil* 34

Chapter 4 *Strength Through Spirituality* 45

Chapter 5 *There's Beauty In The Balance* 52

Chapter 6 *Letting Go* 65

Chapter 7 *The Good Energy* 73

Chapter 8 *Synchronicity* 78

Chapter 9 *Sow Good Seeds* 90

Chapter 10 *A Graceful Mind* 97

Conclusion 108

INTRODUCTION

In this book, I want to introduce you to your third eye chakra, your Ajna, and show you the importance of opening your third eye and what it can do. Chapter One will explain what your third eye really is, and show you how to open it. Chapter Two will reveal the basic techniques and benefits on chakra healing, energy healing, and reiki healing. In Chapter Three, you can find ways to awaken your higher self through guided meditation and how to achieve greater self-knowledge. Chapter Four works towards helping you reach a sense of clarity and wisdom on a higher level than you have experienced. Chapter Five will connect you closer to your intuition and how to balance your chakras. Chapter Six will explain some of the ways you can clear your body of negative energy. Chapter Seven helps you to reenergize your body and mind, a whole new lease on life for you to benefit from. In Chapter Eight, I will use to show you how you can heal your chakras and realign all of your chakras to be in sync with each other. Chapter Nine is all about using the power of your mind to heal yourself from the inside out, reduce your stress, and release your anxiety. Chapter 10 trains you in the power of positive thinking, and purifying your energy field. Throughout the book you will find meditation sessions to practice that I hope you find helpful on this journey. Enjoy!

Only with our mind's eye can we see the truth that is revealed to us.

CHAPTER 1
See The Light

Chakras are energy. More importantly, an energy center working within your body to regulate how it functions. If you understand how an organ such as your pancreas can stop or slow down its production of insulin, the most important hormone required to keep your blood sugars down, and how your doctor might place you on insulin injections or an oral medication to boost your insulin production, then you might understand how an energy breakdown can have a negative effect on your body as well. We begin with the third chakra because of its importance within the chakra system. Do not let the number three fool you. Your third eye chakra, Ajna as it is also known, is tied to the supreme importance of all chakras. By learning how to open this one first, and master that power, this gives you the control you will need if you continue on to opening each chakra.

Ajna is the Sanskrit word for third eye, and translated to English there are two words associated with its meaning. Command and perceiving. By opening the Ajna, you awaken your psychic abilities, your subconsciousness, that greater sense of self that guides your intuition, wisdom, and decisions. Opening your mind to abilities that you have always had but never tapped into can bring many positive changes in your life. I will show you how to stay grounded and in doing so, keep your experiences and the information flowing into you from overwhelming your senses. By incorporating these meditation exercises I will provide you with, along with any others you find helpful, you will be armed with the tools you need to begin this journey and unlock your third chakra, your mind's eye, that sixth sense you've always had but rarely use.

Positioned just above your eye level, between your brows, you can find your third eye chakra, metaphysically speaking. In this same area of your brain are the pituitary gland and the pineal gland. Glands are intricately tied to your chakras, one controls many of the physical functions of your body, and the other commands much of your body's energy.

The chakra system, the one most widely referenced, starts at the base of your spine, with your Root chakra. Just below your navel is your Sacral chakra. In your stomach is the Solar Plexus chakra, followed by the Heart chakra in the center of your chest. The base of your throat is the Throat chakra. Centered near your forehead, between your brows, is your Third Eye chakra, and the one we are focusing on in this series. Lastly, the top of your head is the Crown chakra. You'll see the Sanskrit word for chakras used alongside the English because chakras originate from India and their yoga traditions. Knowing the history of what you are studying can be invaluable at times, so it helps to know where something began.

The exercises you will be reading about next are simple visualization techniques. The mind is powerful, take the time to explore its capabilities and you will be amazed. Don't give up if the results aren't instant. Chances are, they won't be. You just have to give these techniques some time and attention, there will be more than one way to exercise these muscles. You may find one method is easier for you to envision, simply because of where your natural gifts are at this time.

When you begin any psychic or energy work, grounding yourself is an important form of protection. Think in terms of your electrical systems, each one has a grounding wire so as not to blow a fuse.

Just as your electronics need grounding wires, you need a grounding exercise before you begin your own work. I can teach you an easy exercise, the more you use it, the better and stronger you'll become, as with anything you work at.

Since this is your first time with this exercise, take your time, visualize each step, and don't worry about perfection, this is not a requirement. The most important part to this is using your mind to push the negativity out. That is all you are doing. If you find any negativity creeping back into your life, try repeating these steps again. The timeframe is only an estimation, please take as much or as little time as you need, use whatever makes you feel comfortable. When you are ready, find a comfortable quiet spot, turn your phone off so you don't ruin your concentration, and begin.

Ground Yourself (Approximately 3 minutes)

Stand up in your space, tall and relaxed.

Visualize a white magnetic ball of light entering the top of your head.

Slowly move it through your brain, as it moves, the magnet draws all the negative bits of energy from each crevice, and moves slowly into your neck, drawing all the negativity energy you have been holding in your neck.

Move the ball of light through your right shoulder, down your arm to the tips of your fingers and back up again, collecting all the negative bits of energy you held there, and slowly move it across your shoulders to your left, and down your left arm and fingers, again back up.

Take the white ball of light and spread it throughout your back and chest, each piece of negativity attaching itself easily, and releasing tension from your body.

As the ball moves down through your lower body, slowly down through your legs, drawing more of the negative energy from you, you begin to let go of everything you've been holding. As the white ball of magnetic negative energy reaches your feet, you visualize roots beginning to grow and branch out into the earth, you push the energy out through those roots. The earth absorbs the negativity away from you, freeing you from the things holding you back.

Using your mind, push your roots down past the negative energy, until you reach the healing part of the earth's soil. Your roots begin to draw up pure white healing energy up through the rich soil, to the balls of your feet. Now that you are free from negativity energy, you easily attract the pure white energy to you, up through your feet, your lower body, into your chest and back, spreading that warm white healing light through to your arms, across your shoulders, finally up into your neck, and then your head.

Now that you are grounded, protecting yourself is the next step, and even more simple.

Protect Yourself (Approximately 1-2 minutes)

Sit in a comfortable chair with your feet firmly planted on the floor.

Place your hands in your lap, palms up. Close your eyes.

Take 3 slow deep breaths and visualize the following:
Picture a ball of pure, warm, white protective light hovering above your head. As you see it in your mind's eye, it grows bigger and brighter.

Slowly bring the ball down into your head, spreading the light with your mind throughout every part of your head, your neck, down through your shoulders and arms.

Push the protective warm light through your heart, deep into your chest and back, down through your lower body, through to your feet and into your toes. Allow your being to be wrapped and surrounded by this protective white light, shielding you from absorbing any energy but your own.

You've done great work so far, take a rest. Have a glass of ice water, sit back and see how you feel. Are you noticing any tension gone? Bask in the relaxation you are feeling, enjoy the moment. I want the exercises I give you to begin to lead you to a place of inner peace. You cannot rush this journey, it takes time to get there. The more you slow down and work on opening your higher consciousness, you will begin to find a sense of peace enveloping you.

A good way to start this work is to recognize why it is a good idea, and what your intentions are. What would you like to experience? This process will open up your intuition more fully, it has the power to give you a clear focus. Would you like more self-awareness, to be able to handle your stress and anxiety easier? How about your creativity levels, are they high? Or do you find yourself struggling to come up with great ideas? Psychic gifts, such as clairvoyance and clairaudience that give you the skills to see and hear psychically, in your mind's eye, are all benefits to your Ajna being opened.

There are a few different ways you can use to open your Ajna, your third eye chakra. As you read on, together we will explore different ways you can do this, the following method is just the beginning. Depending on how much energy you have blocking your pineal gland, you may have to try this meditation technique several days in a row. A little research will turn up "instant techniques", but you are beginning an exploration into mindfulness, healing, tapping into abilities long held dormant. This takes time and practice. Have patience with yourself and the process. You may only notice a very slight change, the more still you are, the easier it will be to recognize.

Are you ready? By this time you have released all of the negativity you were holding in your body with the grounding exercise, you have surrounded yourself in a powerful white light of protection. Make sure you are warm and comfortable and your phone is turned off. Have a glass of water waiting for you at the end of your session, and more importantly, stay relaxed and open to what may come.

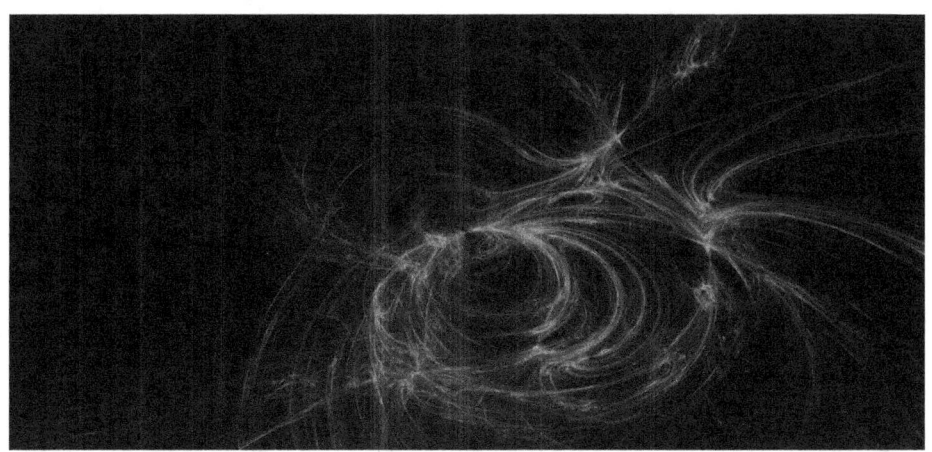

Third Eye Opening ~ The Purple Circle

Start with your breathing, focus on breathing just a little bit deeper than you normally do. Most of us breathe pretty shallow.

Relax your muscles. Close your eyes

Concentrate on the middle of your forehead, removing all other thoughts except your desire to open your Ajna.

Keep an image of your forehead in your mind's eye. Sense your skin, your bones, really feel the presence of your forehead.

Now I want you to see your forehead, with your third eye present in the middle of your brows. Hold this image your mind.

Focus on this picture for a few moments, making sure you can grasp this image in your mind. When you first see it, the eye is closed.

You have the power within you to open it. There is a natural born ability within you, a divine power, you are about to tap into.

Maintain your concentration on your third eye, feeling your positive energy build.

Envision a circle of purple light surrounding the eye.

As you watch, you direct the light to begin opening your third eye chakra, clearing the calcification away.

Stay focused, willing your third eye to open. You may feel pressure in your head, breathe in deeply.

Slowly open your eyes, taking time to notice any different feelings. I can't promise what you will feel or experience, whether it will be instant or over time, and please be very skeptical of anyone who promises you such. What I can promise is there are benefits and psychic gifts to making the effort to accomplish this spiritual awakening. Some people experience an enhanced intuition, or a burst of creativity, some have a lucid dream. A few might notice a very slow awareness coming on. The more you practice opening the eye the closer you become to being a third eye chakra master, and the ways it changes your life will follow. Try this visualization for a few minutes each day, the gifts you will get will be worth the wait.

I shall allow no man to belittle my soul by making me hate him.

— Booker T. Washington

CHAPTER 2
Restoring Your Soul

You've probably heard some people say we have the power to heal ourselves. Seems pretty farfetched, right? How would we even begin to do this? And yet, you know that the mind is powerful, even if you don't know all that it is capable of. You just, know. Yes? Through a lot of therapy, even more talk show therapy, we learned over the years just how much negative thoughts affect our mind, body, even our soul. Any negative thoughts or words spoken get taken in, absorbed, and you begin to believe them, they begin to manifest in your life. This is how abusers keep control over their victims. It does not take much before you begin to believe what is being said, whether by you or someone else. This is why you have to learn to let these thoughts go, daily. The impact on your life, your self esteem, your soul, is staggering, and it is a huge part of what holds you back from achieving any of your goals.

The exercise we did in chapter one to ground you is a great start. You have to make a conscious choice to release destructive and damaging thoughts. An addict who can't shake their addiction is being crippled by this trap in their mind, they do not have the tools to break themselves out of their internal prison. A lot of people judge them and think they're just weak-willed. Contemplate this though. If we all could just will ourselves to wellness we would all be healthy, successful and happy people. Honestly, it is not that simple.

What we can do however, is practice mindfulness, meditation, energy healing, and finding our gifts. We can learn to let go of all the things we hold onto, because in the grand scheme of things, we worry way too much about things that truly do not matter. Think about the moment when a loved one is hurt, but you find out they are going to be okay. Life comes into sharp focus, and you begin to see all that matters is loving each other the best way we can and living a full and happy life. Of course all this is very easy to say, much much harder to do. It is easy for me to write you these words of encouragement. All I ask is that you try some of these techniques, focus on being more mindful and present in the moment, and as it has become a well-known mantra, **DON'T SWEAT THE SMALL STUFF!**

Healing Chakras Meditation (20 – 30 minutes)

Sitting upright comfortably, rest your hands gently in your lap.

Remember, you have the power to heal your being.

With every breath you take in from now on, you bring a pure white light, a divine healing energy that fills your body with a warm, restoring peaceful vitality you have been missing.

With every breath you let out, you release all the worry, tension, and negativity held in your body. Let it go. It is okay.

No harm will come to you.

Only goodness and light can enter your space, your mind, body, and soul

With your eyes closed, take a deep breath, and let out a slow exhale, settling into an easy calmness.

As you exhale, you feel the tension, the negative thoughts leaving as you push out your breath.

Take a deep breathe in, bringing in healing white light into your body.

Exhale slowly, pushing out the worries, bad thoughts, your sadness out, down into the earth to be cleansed.

Breathe in, warm white light that heals with every breath in.

Let your body release every bit of tension you were holding in.

Breathe out, pushing out all the days worries, out and away.

Feel your muscles relax with each healing breath.

Take a breath in, sensing your environment.

Notice the sounds, the smell. The light behind your closed eyelids.

You are becoming one with the space.

Slowly breathe in and out, feeling the rise and fall of your chest.

Let your mind rest, if thoughts come to you, acknowledge them and release them as quickly as they came in.

You have no worries.

You are safe, warm, and secure.

You have no worries.

You are safe, warm, and secure.

Breath down, to where the base of your body rests.

Breath down into your root chakra, below the base of your spine.

With every breath you breath in, you draw into a pure white healing light.

Let your chakra soften, and expand on your breath, giving your root chakra nourishment, and life force energy.

Envision a life sustaining connection between your root chakra and the rich healing soil of the earth.

Picture your root chakra, a glowing red center of light grounding you, strengthening you in the here and now.

Let your root take what it needs.

'I am accepted as I am.'

'I am supported.'

Deep breath in, slowly let it out.

Allow your mind's eye to concentrate on your sacral chakra, just below your belly button, your center of emotional intelligence, choice, creativity, pleasure, and movement.

Feel your breath infuse this area.

Let it soften, and expand on your breath, giving your sacral chakra nourishment and life force energy.

A glowing orange center forms in your mind's eye, the color of the setting sun, balancing, empowering, and motivating you.

'I honor my needs.'

'I allow myself to be nourished.'

Breathe in again. Heal.

Breathe out again. Let go of everything hurting you.

Healing breaths in.

Hurt and worry exhales out.

Move your inner eye up to just below your breastbone, to your solar plexus. This is your chakra of personal power.

Breathe in, softening your solar plexus, envisioning a beautiful yellow light circle, the color of sunshine.

Breathe out.

Bathe your solar plexus in the sunshine, replenishing, restoring, nourishing, letting your chakra take what it needs.

'I value myself.'

'I am enough.'

'I am more than enough.'

Take another breath fully into your belly, and let it all out.

Bring your focus now into the center of your chest, to your heart. Your chakra of self-development and unconditional love.

Gently breathe in deep into your heart.

Let it soften, and expand on your breath.

Invite a bright green light, a beautiful spring color.

Surround your heart in nourishment, renewal, healing, letting it take what it needs.

'I am greatly loved.'

'I allow myself to give and receive love freely.'

'I am nourished by the power of love.'

Take a deep breath in, and out again.

Take your focus up to your neck, your throat, and center on your chakra of personal will and self-expression.

Breathe in, softening and expanding your throat chakra.

Envision a glowing blue circle of light, the color of the sky.

Breathe healing light into your throat center, softening, opening, and freeing your self-expression and creativity.

Let your throat chakra take what it needs. Nourishment, renewal, healing.

'I hear and speak my truth.'

'I express myself freely.'

'I allow myself to have a voice.'

Deep breath in. And out.

Bring your focus up to your forehead, in between your brows, to your third eye.

Your chakra of wisdom and intuition.

Breathing in, you gently allow it to soften, expand, and breathe.

You see a vibrant indigo, the velvety color of the midnight sky.

Bathe your third eye with a circle of indigo, balancing, calming, bringing insight, understanding, lucidity.

Let your third eye take what it needs. Nourishment, renewal, healing.

Everything is unfolding just as it should.

Take another deep breath in.

And out again.

Move up now, to the top of your head, your crown chakra.

Gently, in a warm light softly bathe in your crown. Restoring the balance as you do so, harmonizing.

Let your crown take what it needs from the healing you are providing.

Nourishment, renewal, healing.

'I am one with everything, and everyone.'

'I am one with the universe.'

Breathe in fully.

Exhale fully.

Now.

Feel your full body. As if you were one with your skin.

Feeling the entire outline of your body.

You may have some stored up energy flowing out from you.

With intention and concentration, direct this energy to your belly to be stored there. You can tap into this when you need it.

Before we end this meditation session, take a few moments to acknowledge with love and kindness the unique, beautiful soul you are.

Take one last healing breath in. And you can let it go now.

Slowly open your eyes. Don't stand up just yet. Drink some water. Give yourself a few minutes to come back down to earth.

You have just performed your first chakra healing. I hope you feel better, lighter, more energized and free.

Energy healing is a holistic approach to treating you. Energy is moved to heal you in mind, body, and spirit. It is widely believed if your energy is blocked, this causes illnesses to manifest within you. Think about this. When you neglect routine maintenance on your car, a part may go bad. It starts with a timing belt wearing out, then snapping, and suddenly your valve stroke and piston stroke are smashing into each other. The only thing keeping them apart just broke, now you have bent valves, cylinder head or camshaft damage, and maybe damage to your cylinder wall.

If an energy center in your body breaks down, gets clouded, your energy can no longer flow freely throughout your body. Suddenly you begin to feel rundown and start to get sick. You feel it when you walk around carrying tension throughout your body. Holding bad energy in affects you in ways you cannot see, only feel. An energy healer cleanses that bad energy from you, restoring that free flow of your body's natural energy and rejuvenating you from the inside out.

Reiki healing is just one form of energy healing for your body, mind, and soul. Reiki is not a belief system, and doesn't follow a religion. In its simplest definition, Reiki is an ancient Japanese healing technique of laying on hands. Rei means 'God's Wisdom' or 'Higher Power' and Ki means 'life force energy.' Reiki translates to spiritually guided life force energy. When your Ki is blocked with negative thoughts, or sickness, your life force is low. Your organs and tissues suffer when this occurs, Reiki healing sessions can restore this.

A person learns Reiki through a Reiki master, they are taught to guide God's life giving energy through their hands and into their patient. Light touch, or just above the body are the two techniques used to transmit healing energy. Since it is a divine healing, the energy knows exactly where to go within you to heal you.

Just as with chakra meditation, there are times when the effects felt are very subtle. It could be as simple as an overall feeling of deep relaxation and great stress relief. Others can feel healing immediately, and their complete energy restored. You can benefit greatly from a series of sessions, the benefits cumulate and have a strong effect on your injuries and illnesses.

Chakra healing is a different interpretation than Reiki of energy healing. They have many commonalities and healers use elements of both methods to accomplish their goal, which is restoring you back to full health and wellness.

The main difference between the two practices is simply the energy points in your body. Chakras consist of the 7 major chakras lining your spine straight up to the crown of your head, and minor chakras located throughout your organs and tissues, and even in your energy field. Reiki, on the other hand, simplifies things, and focuses on 3 tandens, main points of energy. Head, heart, stomach.

Since Reiki and Chakra have overlapping points of energy, a healer easily switches back and forth between the two practices, giving them the ability to use what flows easily through them and you to help heal you in the best way possible.

The soul, who is lifted by a very great and yearning desire for the honor of God and the salvation of souls, begins by exercising herself, for a certain space of time, in the ordinary virtues, remaining in the cell of self-knowledge, in order to know better the goodness of God towards her.

– Catherine of Siena

CHAPTER 3
Lifting The Veil

Your soul, your divine spirit, is here. On earth. You are having a human experience. Those mistakes, failures, life struggles you have gone through, sometimes crawled through, they aren't what you think they are. Malcolm Forbes said 'Failure is success if we learn from it.' Each lesson you master brings you closer, more in tune to your soul. Just as your chakras, your energy centers align to make a physically and intellectually happy, healthy human being, aligning with your soul creates a happy, healthy spirit.

Your higher self is that part of you that knows, observes, comprehends everything around you at the most elevated levels you've achieved, while still being present in your human body. Your soul, on the other hand, exists from a different dimension altogether, so the perspective it has is very different than your human self. Every time you view a complicated situation from a different perspective, and realize you're placing too much worry on things that don't really matter in the grand scheme of things, you are evolving. When someone takes a life lesson, goes through it, learns from it, and moves on, they evolve. With each experience, YOU evolve. Don't block the blessings.

Every time you feel a hunch, a flash of intuition, a feeling you weren't sure where it came from, that was your higher self trying to tell you something it wanted you to know. Your higher level of consciousness is the best guide you could have to navigate this human experience we call life.

What gets in the way? You. More specifically, your ego. There is a lot going on in your head, a lot of white noise you can learn to tune out. Your decisions come from a lot of what you have learned. From your parents, teachers, what society tells you is right, or accepted.

Instinctually, you know what is right for you personally. You just don't have a lot of practice listening to your inner voice. And that's why you're here, reading this chapter. I want to show you this connection that is more powerful than you can imagine. We need to nurture that. It's important.

How do you separate your ego from your higher self? This is easy. Your ego identifies with fear, anxiety, tension, anger. It's the conscious and enduring element that learns what it experiences. Your ego takes in all the hurt, disappointment, unhappiness, and negativity you have gone through. That is why carrying that around is so harmful. Ego loves a sense of pride, but shame? Your ego is terrified of that, and consequently will do anything to protect you from that. Shame is really just our fear of being judged and laughed it. It's an uncomfortable place to be. Every time we let go of that, and just take our experience for what it is, good, bad, or ugly, we can take the lesson as our blessing in disguise. We can grow. That is what we are here for. As I said, as many have said before, *don't block the blessings.*

Let's try a small meditation session for this purpose. Awareness of your higher self, that essential connection with your higher self.

The Silk Cord (15 minutes)

Find a comfortable place to lay down.

Begin by laying down and closing your eyes.

I'm going to count back from ten. As I count backwards, slowly feel the tension and negative energy seep out of your skin, down into the floor beneath you, and flow out and away.

10..........9...........8

7...........6...........5

4..........3.............2

1

Your muscles are relaxed, from the top of your head, to your ears, down through your neck, to your shoulders, your arms and hands are relaxed.

Feel the relaxation of your chest, your stomach, and down through your legs to your toes.

Now I want you to place your left hand gently, comfortably, over your heart. This is you, as you are now.

Take your right hand and place it over the break in between your rib cage. This is the seed of your soul, your spirit, your higher self. Make sure you can hold your hands there comfortably.

Breathe in.

Breathe out.

Breathe in a little deeper.

Breathe out.

You are safe, warm, and secure.

Breathe in.

Breathe out.

You are safe, warm, and secure.

Breathe in.

Breathe out.

You are safe, warm, and secure.

You, the physical you, and your soul are infinitely connected. The two of you make up one human being. An amazing human being.

Infinity

Breathe in that thought with that breath.

Breathe out.

Infinity

Breathe in, feeling your heart, your soul connect you. One does not exist without the other.

Breathe out.

Now imagine a white silk cord, a life giving cord, weaving around your left hand over your heart, into a figure eight around your right hand over your soul seed. Trace the infinity symbol around and through your hands.

Breathe in.

Breathe out.

As you take your breaths in and let your breaths out, trace the infinity symbol over and over with your mind's eye.

With each breath in, you feel that connection.

With each breath out, you feel that connection strengthen.

With your third eye chakra, look at the white silk cord wrapped around your hands in the infinity shape.

The white silk cord, showing you how infinitely connected your body and soul are, the cord is glowing.

Your third eye chakra sees that divine white light glowing from the cord as it weaves between your two hands.

Now slowly let your hands lay down by your sides, wherever they are most comfortable for you.

Breathe in.

Breathe out.

The glowing white silk cord is still there.

The infinity cord tying your heart and the seed of your soul is still there.

Trace the infinity shape with your mind's eye.

Feel the bond between the two parts of you that make up one whole being. A divine spiritual being.

I'm going to count back up, when I reach ten, you will be restored, rejuvenated, ready to take on the world with a new awareness of your higher self.

Breathe in.

Breathe out.

1.........2..........3

4.........5..........6

7.........8..........9

10

Open your eyes slowly, keep your breathing even and natural. Slowly, when you are ready, sit up.

How do you feel? You've just used a meditation to see the connection between your body and your soul. That divine connection is unbreakable, and awakening that self-awareness is a great accomplishment.

Spirituality is not a formula, it is not a test. It is a relationship. Spirituality is not about competency; it is about intimacy. Spirituality is not about perfection; it is about connection. The way of the spiritual life begins where we are now, in the mess of our lives.

– *Mike Yaconelli*

CHAPTER 4
Strength Through Spirituality

You are not your thoughts or beliefs. Think about that. What you believed to be true as a child has evolved. You learned many new things in the course of your life so far, and with each new understanding, your thoughts and beliefs reshaped, or changed completely. And yet, you are still you. The core of you is still the same essence, which has not been lost. So the fear that you experience when a long held belief is challenged is not real. You don't have to walk around in fear all the time. That is your ego talking, and your ego will get you into trouble, every time. I'm sure you've heard that many times over. The ego is your path to self-destruction.

When you let go of fear, the need to control the situation, and what you think are your own limitations, you can achieve remarkable things. Even if it's just the mental breakthrough of letting go and just working through it, you've accomplished something great. The confidence you gain to keep going is priceless. There are many times in our lives we are the cause of our own suffering, simply being caught in a mental trap we can't seem to break away from. But we can. The tools are here, use them.

Wisdom is something you carry in you. You know that feeling you get when the hairs on the back of your arms stand up? That sick feeling in your stomach you have when you're in a bad situation? Your body picks up on signals from the world around you constantly, and it is always trying to communicate those to you in feelings and physical impressions. Somewhere along the way you started tuning these out. This is a bad idea. You feel these sensations for a reason. Any guidance you receive from your higher self, your soul even, that is something you want to listen to. They are much wiser than yourself.

Your mind is a great tool for solving your work problems, planning a trip or an event you want to take. Those situations benefit from your intelligence. Exploring your passions, finding ways to be more fulfilled in work and your life, seeking creative inspiration, these situations require your inner voice, that wise higher self-knowledge you carry within your body. When you use the wrong tool for the situation, like your mind to find your passion, that is something that can only be felt, not reasoned. You wind up being mentally blocked, trying to figure out a situation that simply needs the right tool.

So if wisdom is the insightful knowledge you need to make a better decision, clarity is the state of mind, the lucidity you experience when you're in tune with that wisdom.

Sometimes, late at night or early in the morning, you're driving down a road you've been on a hundred times, yet this time you're surrounded by fog. You *know* the landmarks around you, the curves in the road, yet you can't see them, your vision is handicapped by a dense, thick fog in a place entirely familiar to you. You are forced to slow down, approach every bend in the road with caution, until the fog lifts and you can see clearly again. Your psyche becomes enveloped in a type of fog as well. It could be mental, it may be physical, even spiritual fog exists, making it difficult to see your way through.

There are ways to get past this, here are just some of them. Lots of water, exercise, good sleep. I know, boring right? It seems we are always being told we need these. Every single one of these needs helps you stay vibrantly alive, more present. Physical clarity is a great tool to have in your arsenal.

Mental clarity is invaluable. Mental fogs are frustrating, sapping your creativity, and causing you so much confusion. Eating healthy plays a huge role in this. The processed food we consume does so much more damage than you realize. Making fresh food, eating whole foods like fresh fruit and vegetables, lean meats and healthy fats, the benefits to eating this way is invaluable. Not to mention how amazing you feel when you do. The second part of this is letting go of stress. I know, easier said than done, right? It can be a simple start, like a mental pep talk on your way to work to consciously choose to let go of the stress, and then it becomes about forming a habit. We cannot control the world around us. We can only control our own reactions and how we handle the situations that we are in. If you do the best you can, and focus on being the best person that you want to be, it gets easier to be present. Be present, mindful, and do not worry about the way something looks or is perceived. Do not worry about perfection. It's not something we need. Perfection is something we think we want. If you think about it, there is beauty in the flaws. If you don't take the time to stop, and enjoy the experience for what it is, and people for who they are, you are missing out on something incredibly special.

The last one, spiritual fog, usually happens when you're questioning the why of something. When you're asking the universe, your higher power, whoever you pray to, why you, why did this happen to you, that is spiritual fog.

Sometimes I'll hear from a spiritual leader or mentor that everything will work out exactly the way it is supposed to. It is a comforting thought I love to hear when I am going through something painful. Learning that I don't need to regret the choices I made in the past, because they brought me to where I am now, and where I am now is a place I can appreciate, that is a relief to be honest. We can get lost in regret. The truth is, every part of what we have been through is what made us into the person we are today. The more you appreciate that, the easier it is to let regret, worry, and the stress of expectations go. We always say life is too short to live in regret, right? It is not just a saying though. It has a deep meaning to it.

A few ways you can work through a spiritual fog are meditation. You have several meditation exercises here in this guide, and even more can be found on YouTube, in music stores, in your search engine. Meditation is a great way to practice mindfulness, the clarity you will get from that is a huge benefit to you. Positive affirmations are simple, easy tools. Conscious thought is a way to retrain your thinking, to bring about more of what you want in your life. 'I am' is a very powerful statement of personal truth. Whatever words you choose to follow I am becomes your truth. Choose carefully, thoughtfully, with what resonates with you. Write them down and say them aloud to you, put them out for the universe to hear. Give them a chance to manifest in different ways.

The intuition of free will gives us the truth.

– *Corliss Lamont*

CHAPTER 5
There's Beauty In The Balance

Intuition is that unconscious reasoning, that feeling that we need to do something without knowing the why or how of it. In order to hear it, you have to go into a quiet place in your mind. A stillness, turning a listening ear inward to hear where you are being guided. Every time you are listening to your intuition, and connecting with your higher self, you are becoming more in tune with who you are meant to be.

The connection to your intuition is something that everyone has. As we know, not everyone listens. You don't have to dismiss your intelligence, we are striving for a balance between your intellect and intuition. There is beauty in the balance. Take some time for yourself, practice the meditation you are learning, remember to consciously let go of the negativity and worries you carry around with you. They are an extremely heavy burden, and they serve no purpose except to weigh you down.

Our intuition is a quiet voice, especially when we aren't listening. Intuition doesn't involve your ego at all, so all of those egotistical emotions like anger, pride, shame, fear, are gone from that voice. Those emotions are loud, intense, and insistent. If you push past those loud voices, and seek out the quieter, simpler voice free of negative emotion, you have the chance to make a better choice in the heat of the moment. Going with your intuition is being a more authentic version of yourself. Being a more authentic you can never be wrong.

If you want to enhance your intuition, and become a more balanced person, try doing more of the following:

Work creatively. This is a balm to your soul when you pursue your creative side, whether you paint, make a piece of jewelry, put together a scrapbook, write a poem or piece of music. Anything that nurtures that side of you is inspiring, it enhances your intuition.

Keep practicing your mindfulness. The best way to do that is meditation. The second best way is to stop judging yourself. Take your experiences for what they are, but don't judge yourself harshly. Pay attention to it so that you can better understand it. But don't judge it.

Start listening to your gut. Any clues you receive from your body are something worth listening to, so pay attention. Slow down in that moment and try to interpret what your body is trying to tell you.

Strengthen the connection you feel with others. Have you ever noticed how your skin crawls when you see a spider crawl against someone else's skin? When you see the excitement of your favorite team winning, their joy is palpable isn't it? Try observing someone where his or her emotions aren't so obvious. See if you can pick up on what they are feeling just by reaching out with your own senses.

Listen to your dreams. I find it hard to remember my dreams unless I wake up and write them down. When you're unconscious your brain processes things differently, more intuitively, there is a lot of information to be gained from writing your dreams down. They won't make any sense at first. Just write them down, find a good dream interpretation book or website to use. And realize that dream interpretation isn't usually literal. Divine messages some times come through your dreams, by writing them down you have a much better chance of getting the message they want you to have.

Take time for yourself, quiet time to reflect, process through things that have happened, remove the negative thoughts you didn't even realize you had. Keep making this choice, eventually you will find yourself doing this automatically, you'll feel lighter, happier, more joyful.

Your intuition is strong, and the more attention you give to it the clearer it will become.

So why did we begin chapter five with so much talk about your intuition? Your Ajna, that third eye chakra we've been talking so much about, that energy center *IS* your intuition. It is so many things, having it open can lead to clairvoyance, telepathy, lucid dreaming, enhanced creativity and imagination, and visualization.

The five senses, touch, taste, hear, see, and feel, these five senses are what you have relied on since before your birth. Even in the womb you heard your mother's voice, you felt her heartbeat. Your sixth sense, intuition is there for you to rely on as well. You have, I'm sure, at times. Doing this continuously takes practice though.

We find it hard to trust things that we can't see, hear, touch. We constantly search for reassurances that we are making the right decisions. Where do we look? Everywhere except within ourselves and our own instincts. We can fix this though. Like everything else, it just takes practice.

A good way would be to start paying much closer attention to those clues we get. When you work out a deal, it seems great, well intentioned, and then comes the moment when you shake hands with your counterpart, and suddenly the hairs raise on the back of your hand. Your body is telling you *something* is not quite right with this person. The deal goes through, you start to do business, and suddenly you find it's corrupt. Write this down. Every time you feel a physical sense of foreboding, write it down along with the situation, take stock of it from a different perspective. Dig deeper and see if there is something you missed. If you can make better informed decisions by listening closer to your instincts, that intuition you are seeking will get stronger, louder, because finally, you are listening.

We rely on a lot of modern technology nowadays to get us through our day. The ease at which we can accomplish tasks now and manage our lives is pretty amazing. And yet, we are losing touch with relationships that we've had, with ourselves, our family, nature, our spirituality. Everything comes at a price.

I am a trusting person by nature. I look for the good in people first. My sister is the complete opposite of me, she's very suspicious and distrustful. She in turn, thinks I'm too trusting, and she worries I will trust the wrong person. I worry about her instead, always focusing on the negative and expecting the worst first. She worries so much that something bad will happen, that inevitably, when it does occur, she's almost triumphant that she's been proven right all along.

I do my best to trust my own instincts, hoping that I will know when someone is inherently bad, or doesn't have good intentions towards me. Does it always work? No. I have disappointments just like everyone else. I make a conscious choice though, to not focus on the negative attributes of any one person or situation. I know life is a blessing. Our life is a gift we've been given. And if you have people you love in your life that love you back, you're even more blessed. Take the time to enjoy this life, and the people in it. This journey that you're on is fascinating, make the most of it while you are here.

Where were we? Oh yes, balance. I promised you balance.

Let's try an exercise, shall we?
We want to learn how to balance our chakras. The easiest, most direct way to this is meditation. Pay close attention to anything you feel when you try these practices. A physical sensation may be a good sign to show you that it is working.

This is a simple breathing technique that is great for anxiety. It's called *brahmari,* a Sanskrit word that translates to bee. Named for the sound bees make, that low humming sound you hear. Try this in a moment you feel anxious, and see if this helps you. It is also another meditation for opening your third eye. Two birds, one stone.

Brahmari (3-5 minutes)

Sit in a comfortable spot with a tall back, don't forget to relax your shoulders.

Close your eyes.

Start by taking a few slow, natural breaths.

Breathe in.

Breathe out.

Breathe in.

Breathe out.

In. Out.

Place your hands sideways across your face. Your middle fingers go over each eyelid, your index fingers should rest on your eyebrows. Your pinkies naturally settle under your cheekbones. Take your thumbs and close your ears.

Keep your lips tightly sealed. Inhale through your nose.

Exhale with the word AUM, sounding out the M.

Keep sounding out M until you feel the need to breathe in again.

Inhale deeply through your nose.

Exhale with the M sound, as long as it is comfortable to exhale, making the sound as you release your breath.

Inhale deeply, easily.

Exhale as you hum the M sound, humming for as much as your lungs will allow without any discomfort.

Keep doing this for a few minutes until you tire.

When you are ready, return to your normal breathing. Gently. Open your eyes slowly. Take note of how you feel. Hopefully this reduced some of the anxiety you were feeling. Is your tension gone? Do you feel any tingling?

There are other easy methods to balancing your third eye chakra. A simple yoga pose that accomplishes the balance by grounding your third eye to the earth and helps nurture inner eye seeing is the child's pose. Listed below are a few yoga poses recommended for this balancing work.

Child's Pose

Eagle's Pose

Warrior 3 Pose

It doesn't take a lot of strength to hang on.

It takes a lot of strength to let go.

– J.C. Watts

CHAPTER 6
Letting Go

Letting go is something we are all well aware, it's easier said than done. We are all hypocrites when it comes to this, everyone tells everyone else to let things go, and yet we are all holding on to things better left behind. My boss is a mass of contradictions, she tells me one thing, and when I repeat it back the way I heard and interpreted what she said, she immediately tells me she did not say that. I pride myself on being a good listener. Most people in my life tell me what a great listener I am. Almost all that is, except for my boss Judy. This in turn, is a huge source of frustration for both of us, we are always getting our wires crossed. Most of the problem being that she feels I say too much, while I feel she says too little.

I walk into work everyday perfectly fine, the day usually goes smoothly, until she appears. Then I'm instantly tense, because she always misinterprets what I say. Judy happens to think I'm always misinterpreting what *she* says. It's quite comical actually. I constantly struggle between wondering whether I should laugh or cry! This struggle is difficult because it's a daily reality, so unless I feel like looking for another job, I'm stuck with her and she's stuck with me. Can you feel our excitement about that? No? Resignation then. Yes, that's it. So how do I practice what I'm preaching here, and tell you to let go of your daily frustrations, when my own is looming every week, Monday through Friday.

It's true I wasn't handling it well. Judy and I do not speak the same language. Well, we do actually. It's just different dialects. In the Latin culture there are many different Spanish dialects. The Spanish spoken in Spain can be Castilian, Andalusian, or Murcian. There are even more in the smaller regions. Meanwhile the Spanish spoken in the Caribbean is an entirely different Spanish language, they pronounce things very differently and do not use all the same words. They might have originated from the same ancestors, but time, distance, and a natural evolution changed them irrevocably.

So how did I let go? I started listening harder, that's for sure. Sometimes we are too quick to think we understood something, because we assumed certain aspects. I had to ask her to slow down her explanations. Her directions many times are spoken quickly, and then she's off to put out another fire, and I wind up doing the wrong thing because I filled in the blanks myself.

Really, what I concentrate on now is what she really wants from me. Judy wants me to say less. Less is more in her world, and I'm just living in it. I also refuse to let her make me tense anymore. If she misinterprets something she heard me say to a customer, I make her stop and listen. It's not easy when she has her own impression of what I just said, my confidence comes from knowing I'm good at my job and she just needs to trust me.

Letting go is big business. We have songs written about it, therapy sessions devoted to it, a million self help books on it. Proof positive we have a hard time doing so. It is now the year 2018, and we still don't have time machines to go back and undo anything we regret. Do you really want to? Do you want to change the core of who you are today? Whatever occurred that you wish to change, didn't good things come after that, which you appreciated even more? We all have to learn from our past. We are not meant to change it. Only learn. Experience. Move on.

Fears, anger, shame, embarrassment, all of these emotions block you from blessings. As long as you live with and wallow in your ego, you can't evolve, you can't move on to better things in life when you block yourself in. Remember how infinitely tied your heart and your soul are? All of this negativity accumulates in layer upon layer on your body, mind, and soul. Therefore, the process of releasing this negativity is something you have to work at, consciously choosing it each time.

One way you can do this is perform the grounding exercise we tried in the first chapter. Meditation is beneficial to you in so many ways, you just have to try it.

Visualization is another great tool. It helps your mind complete the steps if you picture it first.

The Balloon

It's a beautiful warm day in early summer. You are standing on the bank of a river. The water slowly trickles by, through the canal, over rocks, past you sitting at a base of a tree. You can feel the rough bark against your t-shirt.

In your hand is a balloon, and written in marker is everything you've said you want to let go. Relationships that hurt you, disappointments in your life, anything you could remember to write down. You read over the list, making sure it is complete. If it's not, no matter, you can come back here any time you need to let something go.
Slowly you inhale, letting the string slide up through your fingers.

Now exhale, and let the balloon go, carrying away from you everything negative and harmful you've been holding onto. No longer. Just like that, you've released the bad, so the good things can come back into your life. We can all use more of that.

You can always try the following. Writing anything down is therapeutic, putting words to paper helps you work through whatever you're going through in a really great way. You can sometimes get caught up in a thought process that leads you nowhere fast. Remember that mental fog? I always find by writing something out you learn things you couldn't know before. Talking a situation through is helpful, certainly. Writing though, you process the words differently when you see them in front of you, seeing them in black and white, where you are no longer able to ignore reality.

On a piece of paper, create two columns. Column one is **Everything I Need To Let Go**. Column two is **Who I Need To Tell** (here is where you put anyone affecting you in the situation). Start writing. Do not hold yourself back. Let everything come out on that piece of paper, every thought, every feeling. Regrets, anxiety, the past, whatever you don't like about you. Every time you take responsibility for things you have done, you free yourself from the cycle of creating the same mistake again. In the second column, anyone in the situation with you should be there, including yourself. When you are finished, fold your paper in half.

Begin by writing (or create your own message, but here is a sample): I forgive myself. I forgive (insert names here). Thank you Creator for changing ALL negativity into everything that is light and love.

Speak this out loud, give it voice, and let all of this energy go. And then take your paper and burn it or shred it. Make sure you do this all at once, and be sure to do it alone. This is very personal to you, you don't want to leave this list laying around the house for someone else to find.

When you show deep empathy toward others, their defensive energy goes down, and positive energy replaces it. That's when you can get more creative in solving problems.

– Stephen Covey

CHAPTER 7
The Good Energy

Everything we have talked about in these pages is going to reenergize you. That is, if you try. It all takes work, yes, but anything worth having does, right? If every week you make the goal of incorporating a small change in your life, and give yourself the chance to try it out and create a new habit, you will gain something positive. We all get stuck in a routine that can make life feel very mundane and repetitive. Making changes is invigorating, and I know no one who thinks that that is a bad thing.

On a daily basis, we energize our chakras in different ways. Some of these may surprise you.

Thoughts are powerful. Bad thoughts drain our energy, sap our strength. Positive thoughts keep the energy we have flowing creatively, freely. Where a negative thought will block you and your energy, a good thought frees you, keeping your energy in a positive flow throughout your body.

Tone and sound are vibrations, and they use the same frequency as our organs. Routine toning will help your body's organs working like they are supposed to. Noise pollution is harmful, our bodies find it very disruptive. If you listen to sounds that you find pleasant, this will help you become much more productive and happier as a person.

Color tone is something you can do, just by shining a light through various filters of color and down onto your body, the color vibration gets absorbed into your skin and heals the affected area.

Design is another way you stimulate your chakras. Use colors in an intentional way, color makes you happier, more productive. You find inspiration just by painting a room in a color that makes you more joyful.

Clothing influences your mood, the amount of energy you have, your mind even. Light seeps through your clothing, enhancing the energy of the color we are wearing. Vibrant clothes give you a better transfer of energy than neutral clothes do.

Art and color (as we said) are wonderful outlets to energize you and inspire you. A great idea would be to surround yourself with energizing colors in your workspace and calming colors in your bedroom.

Music affects us in a bad way or a good way. Every note on the music sheet links to a color and a chakra. Certain melodies can evoke a spiritual, mental, physical, or an emotional response in you. When you dance or even just listen to primal music, like beating drums, that can invigorate your body and energize your root chakra.

Aromatherapy uses essential oils, each one matches to a color and have all kinds of healing properties from flowers, herbs, and plants. Make sure the oils you buy are therapeutic, and never apply them directly to your skin. (Follow the directions)

Color bathing is something fun to try. Water is a great energy conductor and color is pure energy. While lying in the bath, try thinking about what you want from that energy center and maybe use the corresponding essential oil. Do not use chemical food coloring to color your bath, there are organic color baths that are safe to use.

Gemstones and minerals carry their own energy. Crystals are structured in their makeup similar to the human body, energy is amplified through them and they can be programmed (they are widely used in watches and computer chips). Having minerals and gemstones in your environment is an easy way to benefit from their healing properties, or simply wear the jewelry.

Visualization is something we've already practiced, several times. Meditation, yoga, and visualization are all thoughtful ways to bring positive energy into your chakra centers. If you add the color intention the vibration becomes even more powerful.

Food grows under the sun's powerful rays, the plants absorbing the sun's life giving energy rays. Everything living gets much needed energy and nourishment from the sun. One way to do this indoors is to get a quality full spectrum light bulb for your living space.

The sun is our most valuable source of energy. It gives life and energy to animals, plants, minerals, water, people, and our chakras. Through the sun's light all seven color energies flow down into the earth.

Happiness is not a matter of intensity but of balance,

order, rhythm and harmony.

– Thomas Merton

CHAPTER 8
Synchronicity

Having a body and mind perfectly in balance with one another is a near impossible achievement. Our lives are messy, complicated, and always throwing us a curveball to deal with. You've probably heard of it though, homeostasis. Our body chemicals, hormones, organs, microorganisms in harmony is a state in which you can thrive, mentally, physically, spiritually, so it is worth pursuing. Balancing your 7 main chakras is an ancient way to get back to this state of being.

Earlier I introduced chakras to you as energy centers. Let me explain further. That energy center is the location where your physical body and your spirit meet. Originally thought to come from Buddhist Tantra teachings as far back as the eighth century, we are focusing here on seven main meeting points from the base of your spine to the crown of your head. The channels between each chakra, 'nadi' as they are called, are the channels through which your life force moves. The Sanskrit word chakra actually means 'wheel' and 'circle'. In your mind's eye picture your chakras as spinning, glowing circles of power, each one their own color, constantly moving your energy throughout your body.

Your root chakra is at the base of your spine, red in color, this is your earth chakra, and gives you feelings of security and safety.

The sacral chakra is just below your belly button, orange in color, this is your water chakra and gives you acceptance, pleasure, and creativeness.

Your solar plexus chakra is where your ribs join over your abdomen, just above your navel. This one is yellow, your fire chakra, it holds your diaphragm and digestion functions, and is where your feelings of control and self-worth stem from.

The heart chakra is at the center of your chest, green in color, and is your air chakra. Your heart chakra gives you feelings of inner peace and love.

Your throat chakra is exactly where you think, in your throat, the color blue, it gives you speech, eating abilities, and taste. This ether chakra represents your communication, internal and external.

Between your eyes is your third eye chakra, the light chakra. Indigo is the color for your third eye and represents your physical sight, your decisions, and intuition.

The crown chakra is just above the crown of your head, shown in the color violet, is your cosmic chakra, and connects you to everything around you, the crown chakra gives you mental peace.

When these are out of line, you can suffer from anger, indecision, jealousy, lack of creativity, addictions, depression, resentment, we could go on for an hour with a list of ailments and conditions that can affect you. You don't even have to believe this is true, wouldn't you just like to feel better? We all want to be healthy and energized in mind, body, and spirit.

This is a simple way to align your chakras and does not take long to do. Just take a few minutes for yourself to try this, see if you're not feeling better by the end. Maybe you'll feel good enough that you'll try it again a second time.

Hand Over Hand Chakra Alignment (10-15 minutes)

Find a comfortable, quiet place to lay down flat on your back.

Lie there for a few minutes quietly, focusing on your breathing.

Inhale. Exhale. Repeat.

Inhale. Exhale. Repeat.

With every breath in, you bring in healing, cleansing energy.

With every breath out you let go of the hurt, physical and emotional, the stress, the negativity you've carried. Let it go.

Let

It

Go

Inhale healing.

Exhale negativity.

Center yourself with your body. Now make your intention known.

You want to align and balance all of your seven main chakras.

Place one hand underneath you at the base of your spine over your root chakra. Place another hand just above your pelvis, below your belly button on your sacral chakra (or just above it, whatever you feel right with).

Keep your hands over each one at the same time. Feel the energy flow and circle between each chakra, with your hands conducting the energy until they align.

See, with your mind's eye, your root chakra and sacral chakra aligning with each other.

Keep your hand over your sacral chakra and move your other hand to your third chakra, the solar plexus, just at the point where your ribs meet.

Now feel your second and third chakra aligning, with your hands conducting the energy stream. Picture this with your mind's eye, remember how powerful visualization is.

Keep your hand over your solar plexus chakra (you can switch hands if it's more comfortable) and place your other hand over your heart chakra.

Synchronize your heart with your solar plexus, let the energy channel freely between each chakra.

Take your time.

Move one hand up to your throat chakra, keeping one hand over your heart, and align them again. Keep the intention present in your mind that they are aligning because you are willing it so. You are in control of your energy.

Leaving one hand over your throat move your free hand to your third eye chakra, over your forehead between your brows. Align these chakras together. It should be getting easier with each one.

Finally move a hand to your crown, leaving your hand over your third eye.

Align these chakras together, using your hands to direct the energy.

Using your mind's eye, trace the line of your chakras from your root to your crown, seeing them glowing together, spinning in the same direction at the same time in perfect alignment with each other.

Breathe deep, inhale, and exhale.

Take a moment or two to rest before getting up.

Yoga is a wonderful exercise that incorporates much of the same principles you find in chakra studies, reiki, energy healing. What yoga does for you is improve your overall health, it naturally balances your energy, heals your everyday aches and pains, builds your muscle strength and flexibility. Your posture gets better with yoga. Practicing yoga gives your muscles and joints a chance to stretch out to their full range of motion, and that helps your joints and cartilage to not break down. Joint cartilage functions just like a sponge, in order to give it fresh nutrients you have to squeeze out the fluid it holds and let new fluid get soaked up. Yoga heals this part of you. It protects your spine, your spinal disks are another part of you that crave movement. Like your cartilage, that is how they get their nutrients.

Yoga is great for bone health and increasing your blood flow and circulation, draining your lymphs. When you practice yoga poses and experience the benefit of draining your lymphs, you're helping fight infection, destroy cancer cells, and rid your body of toxic waste. Upping your heart rate through exercise helps bring down your risk of a heart attack, it helps you fight depression and high blood pressure.

Another benefit is regulating your adrenal glands. They release cortisol in a crisis to boost your immune system. If these levels don't go down soon after, it can permanently change your brain chemistry. Yoga encourages you to eat healthier, it lowers your blood sugar and helps you focus. The relaxation you feel and better balance you gain from the poses are all great benefits. Peace of mind is another benefit, the quiet time you take as you move through each pose gives you a natural calming stillness that quiets your mind.

Yoga connects you to the divine, your higher self, and that is a connection that will help guide you through life, give you a greater sense of self that is priceless. You gain an inner strength with each new thing you try to better yourself, whether it's meditation, visualization, self-healing.

Yoga is an amazing way to practice the mindfulness we have been talking about. The awareness you gain helps you to push through the negative emotions weighing you down, teaching you to let go of the drama in your own life and approach it with a calmer attitude.

We've talked about how much of your body, mind, and spirit overlap, every part of you is intensely interwoven. Every part of this from chakras, to yoga, healthy eating, meditation, visualization, energy healing, all of this embodies a holistic approach to your wellbeing. Every part of this is about taking great care of yourself, self-care is the most important act you can do for you. It just makes you better. Better heart, better body, better mind, better spirit. I know that is always something I am searching for in my own personal life.

Change never comes easy, our daily routines are a comfortable familiar place for us to be. It is what we know. Through this, I hope you'll aim for simple changes, like going to the grocery store and staying in the fresh produce and meats section. Try making one or two nights vegetarian dinners filled with fresh vegetables. Take just fifteen minutes a day and perform a meditation for yourself, see how you feel. Give those small changes a chance to become a comfortable easy habit, and then try another. For example, visit a yoga class. There are many variations of traditional yoga now, you might want to start with the original before you branch off into hot yoga or bikram. Sometimes the simplest answer is the best.

Your minds may now be likened to a garden, which

will,

if neglected, yield only weeds and thistles; but,

if cultivated, will produce the most

beautiful flowers, and the most

delicious fruits.

– Dorothea Dix

CHAPTER 9
Sow Good Seeds

The power contained in your mind is staggering. There it is, untouched, lying in wait for you to realize it's gifts. This is why you have to be so careful with your thoughts, and not let them go unchecked. A lot of them aren't even yours. They're things you've heard, remembered someone saying, read about, and so many have absolutely nothing to do with you. You have heard by now what you focus on is what you attract. If your thoughts are focused on judgment, shame, ridicule, that is what you are bringing to yourself. I'm sure you've heard that what a bully picks on in another person is really an insecurity they have within themselves. Makes sense right? His or her jealousy is on display for everyone to see.

What you need to remember is that as long as you hang on to negative thoughts, about yourself, someone you love, or even someone you hate, you've left no room for positive thoughts. You can't have both at the same time, negative is always going to dominate the positive. Your pessimistic side is louder, more insistent, and likes to take over. Your job is not to let that happen.

The human mind is a garden of fertile soil, what seeds you sow are what will grow. Your subconscious does not judge, it provides what you plant for it. You can ask, through repetition and feeling, for all things good in your life. This has to be a conscious choice. That is why so many recommend trying positive affirmations as a way to create this in your life.

Using your thoughts, images, your voice to bring what you need into your life, practice this daily, and watch a synchronicity unfold. Conscious thought and choice will bring you to the places and people you need to achieve your goals.

You've seen how everything about you and your life has energy, from each organ and tissue, to your mind, the colors you see, the gemstones you wear. Now that you know your thoughts have this same energy, visualization, affirmations, beliefs, desires, your fears all have a huge effect on your reality. Each one has their own energy vibration that reverberates throughout our lives and into a web of reality. Keeping us all connected to each other.

When you say something and then repeat it to yourself, this has a huge influence on your thoughts. So it is very important to keep these positive, simple, and short. Repetition is the key if you want to see it work for you.

Successful people have the benefit of momentum, they already have the law of attraction working for them and if they keep their thoughts positive, they attract greater success. You don't have to have their same level or attributes to have this work for you, start with acknowledging how you are already successful. Take everything you think is good about you, all of your positive qualities, and make a list. You can target this to a goal you have or just your life in general, and this creates a successful vibration.

Getting rid of negative thoughts is not as easy, we've done a meditation on this and even talked about consciously letting go, but it is still hard. Anxiety creeps in and suddenly your bad thoughts are multiplying like bunnies. The next time you are worried about what might happen, envision everything that could go right. You can only think one thought at a time so by choosing positive thought, you solve the problem easily with only a change in thought pattern.

The key to this is stopping the momentum of bad thoughts. You can choose not to give them power. The more you notice pessimistic thoughts and try to change them, the process becomes easier to do each time, before you know it you've created a great new habit that leaves you happier, less stressed.

Anxiety is not so easy to overcome, if it were it wouldn't be a billion dollar industry. When you're in the middle of an anxiety attack, so much of that is caught up in your emotions, a thought change will not stop it.

The first thing to do is to sit and focus on your breathing. Keeping your eyes closed, rest one hand on your chest, over your heart chakra, and one hand over your sacral chakra. Breathe in, breathe out. As your thoughts start to overwhelm you again, put the focus back to your breathing. Keep breathing gently, consistently, until you feel calmer.

Soda is a drink that depletes vitamins and minerals from your body and strips the enamel from your teeth. Not only will you be saving money by eliminating it from your shopping list, you'll lose weight and save money on your dental bill.

Trim the fat from your budget and track where you are spending your money. The anxiety you experience from financial stress is something you can take control of, if you plan carefully.

Clear out the clutter. If you've ever watched hoarders you've seen that accumulating stuff can put your health, both physical and mental, at great risk.

Get more sleep and exercise. A restful night's sleep and any form of exercise do so many things for you, and reducing anxiety is just one of those benefits.

Wake up fifteen minutes early. Rushing around in the morning contributes greatly to your anxiety, give yourself more time in the morning so you can relax, take a deep breath, and start out positive.

This tip goes hand in hand with getting rid of the soda. Eliminate sugar, processed foods, and reduce your caffeine. These can actually bring on anxiety and panic attacks, even more so if you suffer from a disorder.

Finally, one of the best ways to reduce your anxiety is to schedule a visit with a good therapist. Do some research, read their bios, ask for recommendations, and if you don't feel comfortable with them, try someone else. There is nothing wrong with finding a therapist you click with, if you don't make the effort you most likely won't get anything out of it.

Give yourself a gift of five minutes of contemplation in awe of everything you see around you. Go outside and turn your attention to the many miracles around you. This five-minute-a-day regimen of appreciation and gratitude will help you to focus your life in awe.

– Wayne Dyer

CHAPTER 10
A Graceful Mind

I wanted to end this book with purifying your energy field. With all of the energies we carry within us and come into contact with everyday it can be very overwhelming and you don't even know it. When your emotions are all over the place and you feel anxious, you can try a simple practice to purify your own energy field and see if that brings any balance back to your life.

Baking soda and sea salt are two cleansers that are easy to find almost anywhere. Salt water is very powerful, place some in your shower and try this simple practice. Pour salt into your hands and move them over each chakra, asking each one to balance. If you picked up anyone's energy you can picture them and send a simple prayer of love and light. Now you can say out loud or to yourself 'With this salt I purify myself in body, mind, and spirit. I release anything bad and anything that isn't mine.'

If that does not leave you feeling cleansed, take a bath with baking soda and practice the same thought technique.

Apricot seed oil or coconut oil lotion can be used as a form of personal protection between you and the world. If you have a small quartz crystal you can place that inside and set your intention of this being a protection for you.

Sage your space. Sage sticks, when lit, release a smoke into the air and this clears out all the negative emotions and vibes lingering behind. If you've just had a fight with someone perhaps, or find everyone in your home are feeling tense and unhappy, saging is a great way to clear that emotion out and start with a clean slate.

Clean your home. Believe it or not, this also purifies your energy field. You know the saying, clean home, clean mind, right? It's true. Get rid of the clutter. Change the energy of your space.

Dance. Remember how music and dance reenergizes your chakras? It clears your energy field as well.

Picture your energy field, surrounding you, protecting you. Envision a circle of energy surrounding you about four feet away. Bring the circle one foot closer. And again, so that it's just two feet away. Now one foot. Once more, bring that energy field closer to you, until it is against your skin. Draw the energy into your core now. Hopefully you feel safer, reinvigorated.

More importantly, don't pick up someone else's energy. It doesn't belong to you. If you notice you have done that, imagine a rosebud, slowly blooming as it picks up all the energy that is not yours. Once the rose is full, send it out, into the air, watching it burn in a light of purple fire, purifying as dissolves.

Cleansing is something that can be done physically, spiritually, mentally, as I've shown you here. Notice that almost everything involved is free, but for those few exceptions.

Nature is probably the best way to renew you in mind, body, and spirit. Mother Nature is our greatest energy healer, and yet we take her for granted. Find a park with trails and gardens to walk through. Just by crouching down and placing your hands on the earth, visualize letting go of anything weighing you down and letting the ground absorb it all from you. Take your time walking through the trails, absorbing the stillness of the trees. Find a spot to sit with your back against a tree and take the time to write, draw, paint, or even listen to beautiful music. Before you leave you will find yourself renewed, I promise you.

All the work we have done so far to learn meditation and visualization, the health benefits to positive changes in your life, and the energy work has the goal of making your life better. I want you to tap into your gifts and this is why we worked on opening your third eye chakra. When it does open, your intuition gets stronger, easier to follow. Emotions become easier to balance and reason. You can gain clarity in the beginning, suddenly seeing things in a clear mind, and then go back to a period of confusion. Keep your thoughts positive, the image in your mind of a clear and open third eye and the momentary cloudiness should disappear.

You may start hearing your mind, as if it was one person talking to another. Your mind is just talking to itself, until now a lot of thoughts went undetected, if you suddenly start hearing them this can be disconcerting. It will become very important to notice when your mind is being irrational, make sure you don't let that overwhelm you. Keep in mind the gifts you receive from opening your eye are different than someone else's. Your experience will be different, and will happen in it's own time. Of course the more you put into the practice, the more you will receive from it.

When you first open your eye, the colors you see and images shown aren't clear. With patience, and practicing your visualization and meditation, all of this will begin to clear.

Here's a way you can see if your third eye is awakened. Close your eyes. Can you see:

- Purple/white/blue colors
- Intense white dots
- Black sky filled with stars
- A circle/square/eye/any shape filled with purple or blue color

Any pressure you feel, or movement is a great indication your third eye is waking and soon you can see with it.

You may become sensitive to light, and see colors vividly. Headaches may increase as well. The information can overload you. Take a walk outside, or do a small meditation exercise. Drink more water.

You may feel a slow change, more sharpened senses. Embrace those changes. Wear indigo and purple. Daydream.

If nothing has happened and some time has passed, make sure to keep doing the meditation. You can try the following foods as well:

- Foods rich in omega -3 like sardines, salmon, chia seeds, and walnuts clear your cognitive function and will help open your third eye.
- Dark chocolate enhances mental clarity and boosts your concentration. The magnesium in chocolate destresses you.
- Purple foods promote balance in your third eye so eating eggplant, purple cabbage, blackberries, blueberries, and red grapes all work to help you unblock your third eye chakra.

There are stones that help you with this goal, these three are the ones you would one to use.

- Black Obsidian gives a balance between reason and emotion.
- Purple Fluorite is a semi-precious gem that helps to sharpen your intuition and clear your confused thoughts. You can use this crystal when you're needing to make a hard choice and need to rid yourself of distractions.
- Amethyst is a purple precious stone that promotes healing and wisdom. Amethyst also helps with headaches (a side effect of a blocked third eye).

If you are experiencing sensations or seeing images and colors, here is more of what you can expect if you continue this practice.

As you lay down, tired and ready for sleep, images may fill your mind's eye. Most won't make any sense, and don't be disappointed if they're blurry. It takes time to develop this sight, you can't run before you crawl, right?

Your abilities are attuned to your own unique vibration. The lower your vibration, the weaker your spiritual abilities are. The best way to raise your vibration and make the images and visions clearer is to embody gratitude, happiness, peace. Eat fresh and healthy as often as possible, do some of the suggestions I wrote here. Anything that brings you lots of peace and happiness personally will naturally raise your energy and vibration and keep your chakras clear.

If you have your eyes closed, and see a vision that scares you, just remember to open your eyes, maybe turn on your favorite calming music, and do not be afraid of it.

Auras may be visible to you. If you'd like to see them, you can try concentrating on one person and picturing their aura around them, like a rainbow around their head. What colors can you see? Make sure you use your third eye and try to detect how many colors surround them.

Your dreams become more vivid with your third eye being open. Lucid dreaming is a sleeping meditation. It feels real and vivid, and having them is wonderful for your creativity. Do not be afraid of your nightmares, they are a much needed release for your negative emotions and thoughts. Nightmares are healthy, surprisingly enough.

Shadows may suddenly be visible to you. We exist in the 2nd dimension, many spirits live in the 3rd dimension, and if you want to, you can use your third eye to see this. If this scares you, you can close it, just understand you may be closing all your gifts that your third eye brings, and not just the gift of sight.

Prescience is the gift of seeing something before it happens. This is a psychic ability, and usually works hand in hand with your highly developed intuition.

Truth seekers are people usually searching for more than just what they've been told. Your third eye chakra helps you discern illusion from truth.

It is my personal belief that we all have spirit guides and guardian angels, helping us without us ever knowing.

If any part of opening your third eye causes you any stress or discomfort, you can ask your guide to slow them down so that you may better understand them. You don't even have to believe they are there, just know that you can ask for help at any point in time. I believe if you ask, that help will come to you.

You can always ask for more guidance and protection, I firmly believe if you ask you shall receive. You can also do the grounding and protection exercises again, especially if there has been some time between, then you may need to do so.

One last technique to help you open your Ajna and protect you is the simplest one yet. Place a beautiful glowing amethyst stone just over your third eye. Envision what that looks like. Doing this will not only help you open your Ajna, amethyst will also protect you from the visions that may flood into your consciousness. They can be overwhelming, so any time you do anything to ground yourself it is a good thing.

The more relaxed you are for all parts of this, the better you will receive the information about to come to you. Leave your impatience behind, remember that's an ego trait, just open your mind to receiving the information gently and peacefully, and it will come. Enjoy the journey!

CONCLUSION

Thank you for making it through to the end of this book. I hope it was informative and able to provide you with all of the tools you need to achieve your goals, whatever they may be. The next step is to start trying some of these techniques in your own life, and find out what works best for you. Lastly, if you enjoyed this book I ask that you please take the time to rate it on Amazon. Your honest review would be greatly appreciated. Thank you!

www.ingramcontent.com/pod-product-compliance
Lightning Source LLC
Chambersburg PA
CBHW071501070526
44578CB00001B/404